ÖHNE

A. LANGE & SÖHNE®
HIGHLIGHTS

HENNING MÜTZLITZ

4880 Lower Valley Road • Atglen, PA 19310

Other Schiffer Books on Related Subjects:

American Wristwatches: Five Decades of Style and Design. Edward Faber & Stewart Unger, with Ettagale Blauer. ISBN: 0-7643-0171-3. $79.95

Automatic Wristwatches from Germany, England, France, Japan, Russia and the USA. Heinz Hampel. ISBN: 0-7643-0379-1. $79.95

Breitling Highlights. Henning Mützlitz. ISBN: 978-0-7643-4211-0. $29.99

Breitling: The History of a Great Brand of Watches 1884 to the Present. Benno Richter. ISBN: 978-0-7643-2670-7. $49.95

Omega Designs: Feast for the Eyes. Anton Kreuzer. ISBN: 0-7643-0058-X. $59.95

Omega Highlights. Henning Mützlitz. ISBN: 978-0-7643-4212-7. $29.99

Rolex Wristwatches: An Unauthorized History. 2nd Edition. James M. Dowling & Jeffrey P. Hess. ISBN: 0-7643-1367-3. $125.00

Library of Congress Control Number: 2013940164

A. Lange & Söhne Highlights written by Henning Mützlitz was originally published by HEEL Verlag GmbH.

This book was translated by Elizabeth Doerr.

Cover by Justin Watkinson
Type set in Tall Films Expanded/ITC Avant Garde Gothic Std

ISBN: 978-0-7643-4361-2
Printed in China

Published by Schiffer Publishing, Ltd.
4880 Lower Valley Road
Atglen, PA 19310
Phone: (610) 593-1777; Fax: (610) 593-2002
E-mail: Info@schifferbooks.com

For our complete selection of fine books on this and related subjects, please visit our website at **www.schifferbooks.com**. You may also write for a free catalog.

This book may be purchased from the publisher. Please try your bookstore first.

We are always looking for people to write books on new and related subjects. If you have an idea for a book, please contact us at proposals@schifferbooks.com

Schiffer Publishing's titles are available at special discounts for bulk purchases for sales promotions or premiums. Special editions, including personalized covers, corporate imprints, and excerpts can be created in large quantities for special needs. For more information, contact the publisher.

In Europe, Schiffer books are distributed by
Bushwood Books
6 Marksbury Ave.
Kew Gardens
Surrey TW9 4JF England
Phone: 44 (0) 20 8392 8585; Fax: 44 (0) 20 8392 9876
E-mail: info@bushwoodbooks.co.uk
Website: www.bushwoodbooks.co.uk

A.LANGE & SÖHNE

CONTENTS

1 .... **Brand history** .... Page 6

2 .... **Historical Models** .... Page 16
A selection of the most well-known wristwatches of the 1920s, '30s, and '40s

3 .... **Highlights** .... Page 26
The most spectacular wristwatches from the collections of the last 15 years

4 .... **Lange 1** .... Page 40
This modern classic has been the "face" of A. Lange & Söhne since 1994

5 .... **Richard Lange** .... Page 54
The homage to the company founder's son is characterized by great precision

6 .... **1815** .... Page 58
The year of Ferdinand Adolph Lange's birth is behind the mysterious name of this line

7 .... **Cabaret** .... Page 66
This perfect rectangle also appeals to the ladies

8 .... **Saxonia** .... Page 74
The name of this alternative to the Lange 1 models clearly pays homage to its homeland

9 .... **Langematik** .... Page 80
No more winding: a typically classic A. Lange & Söhne outfitted with automatic winding

10 .... **Arkade** .... Page 88
Inspired by the Dresden Castle, the Arkade is always a great addition to any female wrist

A. LANGE & SÖHNE

PREFACE

A. Lange & Söhne has made history during the last fifteen years: Saxon history, German history, economic history, and—last but not least—horological history. By connecting with its own Erzgebirge tradition, the Glashütte brand has once again become more than the nostalgic memory the socialist era made it into: A. Lange & Söhne has now become an animated leader in the world of watchmaking—from Saxony.

Alongside the brand's extensive history, this book presents you with about one hundred of the most beautiful and sought-after models manufactured by the Glashütte-based marque. In this volume you will not only find pre-war and wartime models sought after on the secondary market, you will also get to know the grand highlights of more recent brand history that stand in the spotlight. Timepieces such as the Tourbograph Pour le Mérite are among the most spectacular in the history of mechanical watchmaking and allow the observer to dream. Modern classics such as the Lange 1 and the 1815 models have resuscitated the image of both the brand and its location Glashütte in the last decade and a half.

The selection of timepieces shown in this volume represents only a cross section of the entire palette of variations that the brand has manufactured since 1994. It is without a doubt just the right amount of images for browsing, and these will certainly ignite your curiousity to see more of the complicated, innovative watchmaking this illustrious Saxon brand is highly capable of producing.

Heidelberg, January 2010
Henning Mützlitz

Ferdinand Adolph Lange

The traditional Lange residence

A. LANGE & SÖHNE

STATE-OF-THE-ART TRADITION

The history of A. Lange & Söhne began in 1815: pretty much the same year that Europe—still in ruins thanks to the French Revolution and the Napoleonic Wars—received a new order at the Congress of Vienna. After years of conflicts, its settlements brought a stability able to create a platform for economic upturn in Europe in the ensuing decades despite a reestablishment of traditional modes of thought. On February 18 of that same year, Ferdinand Adolph Lange was born the son of gunsmith Johann Samuel Lange in Dresden. At the ripe old age of 15, in 1830, still in the midst of his schooling, he began an apprenticeship with later watchmaker to the royalty Johann Christian Friedrich Gutkaes. By the time 1835 rolled around, he had completed his apprenticeship, remaining another two years in the workshop of Dresden's most renowned watch business. When the young watchmaker learned of the flourishing centers of watchmaking in London and Paris, he began his journeyman years in earnest. He worked and studied for four years in Paris with chronometer maker Thaddäus Winnerl to learn thc secrets of high watchmaking and, more precisely, to penetrate the scientific side of his profession. Many meticulous ideas and inventions by master watchmakers Abraham-Louis Breguet and George Graham thus found their way into his journeyman notebook.

In 1841, Lange returned to his Saxon home and one year later married Antonia Gutkaes, his previous master's daughter. The very same year he became a master watchmaker and a partner to Gutkaes,

The monument honoring Ferdinand Adolph Lange

Johann Christian Friedrich Gutkaes

who had been called to the royal court as its watchmaker. The young watchmaker Carl Moritz Grossmann began an apprenticeship with Gutkaes that same year.

At first, Lange designed several complicated timepieces for Gutkaes & Lange, initiating a remarkable economic upswing for his company. Thus inspired, he submitted an offer to privy councilor von Weissenbach to build a pocket watch factory in the structurally and economically weak Erzgebirge region, in the mountains surrounding Dresden.

With the beginning industrialization, and particularly the expansion of the railroads even in Saxony, there was an increased demand for precision watches and pocket chronometers to synchronize timetables. In this way, Lange wanted to allow the entire region to profit from the rise of mechanical watchmaking, of which the Saxon ministry of the interior was particularly fond. The concept of the young entrepreneur included the manufacture of one single type of watch using standardized movement design, purpose-built machines, and an organized division of labor. The apprentices were to originate in the region and be able to complete twelve watches per week, which would result in an annual production of 600 pieces. He also thought to have some craftsmen make cases in order to attain more independence from foreign suppliers.

This was also true of the movement design itself. By developing the three-quarter plate and the Glashütte lever escapement, Lange created prerequisites for pocket watches designed to meet the challenges of a new era.

The Saxon government agreed with his plan, and in 1845 the Glashütte-based manufactory Lange & Cie. was founded. After a few difficult beginning years, in which few of the planned 600 watches could be completed, Glashütte bloomed as a location for watch

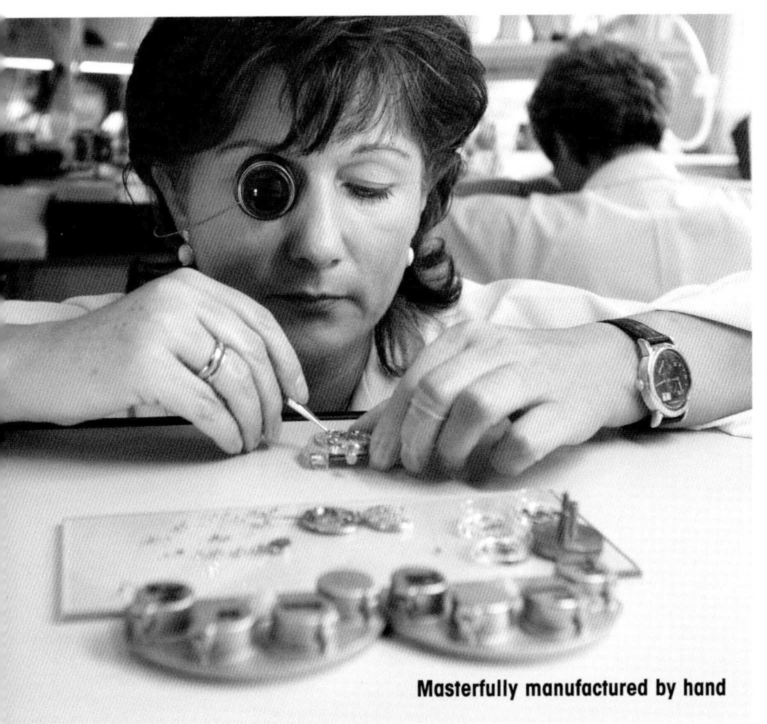
Masterfully manufactured by hand

Lange Uhren GmbH's personnel

manufacture: Lange's first students were already able to gain their independence in 1848 and founded their own workshops for supplier components in and around Glashütte. Other melodious names that were to become legendary in watchmaking worked alongside Lange in Glashütte: Friedrich August Adolf Schneider, Julius Assmann, and Gutkaes's apprentice Carl Moritz Grossmann.

A few days after founding the company, Lange's son Richard entered the world. Like his father he was trained as a watchmaker, after which he journeyed, and advanced, to become a master of his trade. Ferdinand Adolph Lange also acted as the mayor of Glashütte from 1848 until 1866, in addition to leading his company.

In 1868, Richard joined the company, which then changed its name to A. Lange & Söhne. Ferdinand Adolph's second son, Emil, followed a few years later. In 1873 the Lange residence was built, which served both as workshop and family home.

On December 3, 1875, Ferdinand Adolph Lange passed away, after which his sons jointly continued his work. Twenty years later, on the occasion of the fiftieth anniversary of the company's founding, the city of Glashütte erected a monument to the pioneer in watchmaking that can still be admired in the town.

In the subsequent period, Lange's sons seamlessly continued the work of their father, and Emil received the Knight's Cross of the French Legion of Honor in 1902 for his service in watchmaking. Four years later, the third generation entered the company in the form of Emil's son Otto. His other two sons, Rudolf and Gerhard, also joined the management, and from 1919 the three brothers successfully continued leading the company.

Richard Lange discovered a new beryllium alloy in 1930 that improved the characteristics of springs in watches and patented this invention. A new movement line called Caliber 48 was manufactured from 1938, which was used during the war years in the pilot's and marine observation watches of the German armed forces. On May 8, 1945—the day that Germany officially capitulated—an air raid destroyed the company's main production building. The production of Caliber 48

In 1945 the Lange factory was in ruins

was only continued after that with great effort. In 1948—in a divided Germany—the company was confiscated and the family expropriated. A. Lange & Söhne was turned into a "people's company" called VEB Mechanik Lange & Söhne. In 1951 the company, like the entire Glashütte industry, was integrated into a combine, thus becoming part of Glashütter Uhrenbetriebe, which remained in place until the Berlin Wall fell in 1989.

Walter Lange, born in 1924, the great-grandson of Ferdinand Adolph Lange and trained as a watchmaker before World War II broke out, took the unexpected opportunity after German reunification to refound his family's legacy from the ruins of socialism. On December

Modern precision for a traditional product

7, 1990—precisely 145 years after his great-grandfather originally founded the company—he registered Lange Uhren GmbH in Glashütte and globally protected the brand name A. Lange & Söhne. This act marked the beginning of a new era in the company's history.
The first manufacturing space of the new era was housed in the building of former precision pendulum clock company Strasser & Rohde. From there, the firm produced a debut collection in October 1994, firmly positioning the returning brand in the spotlight of the world's stage of fine watchmaking in the modern age: Lange 1, Arkade, Saxonia, and the Pour le Mérite tourbillon sealed a new reputation for A. Lange & Söhne.

Walter Lange

The renovated traditional Lange residence

LANGE
A. LANGE & SÖHNE
25

Just one short year later, the Lange 1 model won readers' hearts as "Watch of the Year" of the German special interest magazine ARMBAN**DUHREN**. At this book's press time, timepieces by A. Lange & Söhne had won this particular award a total of seven times out of sixteen.

In 1995, Walter Lange became a citizen of honor of the city of Glashütte. Together with managing director Günter Blümlein, Lange led the brand back into the top echelon of global watchmaking, while its global reputation and production numbers continuously increased. In 1998, the brand inaugurated a second production building in Glashütte. Two years after that, A. Lange & Söhne was integrated into the Richemont Luxury Group.

The new millennium rang in even more production expansion and global success: in 2001 the former Lange residence was renovated and put once again to use as a place of production and as a schooling center for apprentices. A new "technology center" followed two years after that, which is also home to the in-house production of balance springs. In 2007, the very first A. Lange & Söhne mono brand boutique was opened in Dresden, near the refurbished Frauenkirche. Two more have since followed, located in Tokyo and Shanghai respectively.

A. Lange & Söhne is without a doubt once again positioned at the top of the global horological pyramid—completely in the tradition of its legendary founder Ferdinand Adolph Lange and his descendants.

HISTORICAL MODELS

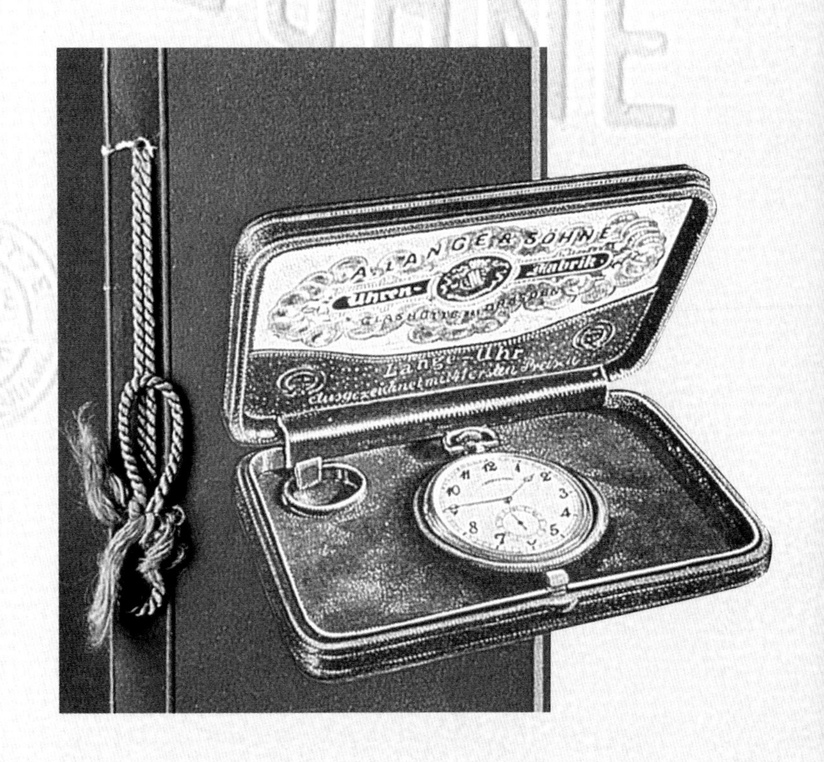

Lange has manufactured wristwatches since the 1920s, above all outfitted with Swiss ébauches. During the Third Reich era and World War II, the Saxon company chiefly manufactured pilot's and observation watches for the German Luftwaffe.
Case diameters of 55 millimeters [2.1"] were practically standard, and some observation watches even reached diameters of 65 mm [2.5"]. Alongside these, Lange also manufactured marine chronometers. The standard chronometer produced in conjunction with Hamburg Chronometerwerke achieved a great deal of fame due to its precision; a rate deviation of just one second every three days would still be sensational, even in this day and age.

MEN'S WATCH (1940)

Movement: manual winding, rhodium-plated, côtes de Genève
Functions: hours, minutes, subsidiary seconds
Case: yellow gold, ø 35 mm; push-down case back
Estimated value: (2009) €6000,– [$7700]

MEN'S WATCH (1944)

Movement: manual winding, Caliber 10 1/2 ''';
nickel-plated, côtes de Genève
Functions: hours, minutes, subsidiary seconds
Case: yellow gold, ø 36 mm; push-down case back
Estimated value: (2009) €6000,– [$7700]

TECHNICAL DATA

MEN'S WATCH (1942)

Movement: manual winding, Caliber 10 1/2 ''';
rhodium-plated, côtes de Genève
Functions: hours, minutes, subsidiary seconds
Case: yellow gold, ø 36 mm; push-down case back
Estimated value: (2009) € 6000,– [$7700]

MEN'S WATCH WITH LANGE CALIBER (1939)

Reference number: 2562
Movement: manual winding, Lange Caliber 31; gold-plated, frosted finish, jewels set in chatons
Functions: hours, minutes, subsidiary seconds
Case: yellow gold, ø 35 mm; push-down case back
Estimated value: (2009) € 15.000,– [$19,200]

MEN'S WATCH (1928)

Movement: manual winding, rhodium-plated, côtes de Genève
Functions: hours, minutes, subsidiary seconds
Case: yellow gold, 38 x 23 mm; push-down case back
Estimated value: (2009) € 6000,– [$7700]

MEN'S WATCH (1932)

Movement: manual winding, gold-plated, frosted finish
Functions: hours, minutes, subsidiary seconds
Case: yellow gold, 38 x 22 mm; push-down case back
Estimated value: (2009) € 2500,– [$3200]

TECHNICAL DATA

MEN'S WATCH (1940)

Movement: manual winding, Caliber 8.3/4x12''';
rhodium-plated, côtes de Genève
Functions: hours, minutes, subsidiary seconds
Case: yellow gold, 36 x 21 mm; push-down case back
Estimated value: (2009) € 2500,– [$3200]

MEN'S WATCH (1936)

Movement: manual winding, Caliber 8 3/4''';
nickel-plated, côtes de Genève
Functions: hours, minutes, subsidiary seconds
Case: yellow gold, 38 x 21 mm; push-down case back
Estimated value: (2009) € 2500,– [$3200]

OBSERVATION WATCH FOR THE GERMAN WAFFEN-SS (1940)

Movement: manual winding, two-third base plate, gold-plated, frosted finish, jewels set in chatons
Functions: hours, minutes, sweep seconds
Case: silver, ø 65 mm; push-down case back
Estimated value: (2009) € 25.000,– [$32,100]

OBSERVATION WATCH FOR THE GERMAN AIR FORCE (1936)

Movement: manual winding, Glashütte gold pallet lever and escape wheel, frosted finish, gold-plated, jewels set in chatons
Functions: hours, minutes, sweep seconds; gauge
Case: nickel, ø 58 mm, push-down case back
Estimated value: (2009) € 20.000,– [25,700]

TECHNICAL DATA

PILOT'S WATCH WITH SPECIAL DIAL (1938)

Movement: manual winding, Lange Caliber 45;
Glashütte gold pallet lever, gold-plated, frosted finish
Functions: hours, minutes, sweep seconds
Case: silver, ø 55 mm, push-down case back
Estimated value: (2009) € 20.000,– [$25,700]

GAUGE OBSERVATION WATCH FOR THE GERMAN AIR FORCE (1939)

Movement: manual winding, Glashütte gold pallet lever, gold-plated, frosted finish
Functions: hours, minutes, sweep seconds, gauge
Case: silver, ø 55 mm; push-down case back
Estimated value: (2009) € 20.000,– [$25,700]

PILOT'S WATCH (1940)

Movement: manual winding, Lange Caliber 48.1; gold-plated, frosted finish
Functions: hours, minutes, sweep seconds
Case: grey matte stainless steel, ø 55 mm; push-down case back
Estimated value: (2009) € 4500,– [$5700]

PILOT'S WATCH (1940)

Movement: manual winding, Lange Caliber 45; Glashütter gold pallet lever, gold-plated, frosted finish
Functions: hour, minutes, sweep seconds
Case: grey matte stainless steel, ø 55 mm; push-down case back
Estimated value: (2009) € 12.000,– [$15,400]

TECHNICAL DATA

FIGHTER PILOT'S WATCH FOR HERMANN GOERING (1941)

Movement: manual winding, gold-plated, frosted finish
Functions: hours, minutes, sweep seconds
Case: silver, ø 55 mm; push-down case back
Estimated value: € 40.000,– [$51,400]
(2009)

PILOT'S WATCH (1943)

Movement: manual winding, gold-plated, frosted finish
Functions: hours, minutes, sweep seconds
Case: lacquered stainless steel, ø 55 mm; push-down case back
Estimated value: € 4500,– [$5700]
(2009)

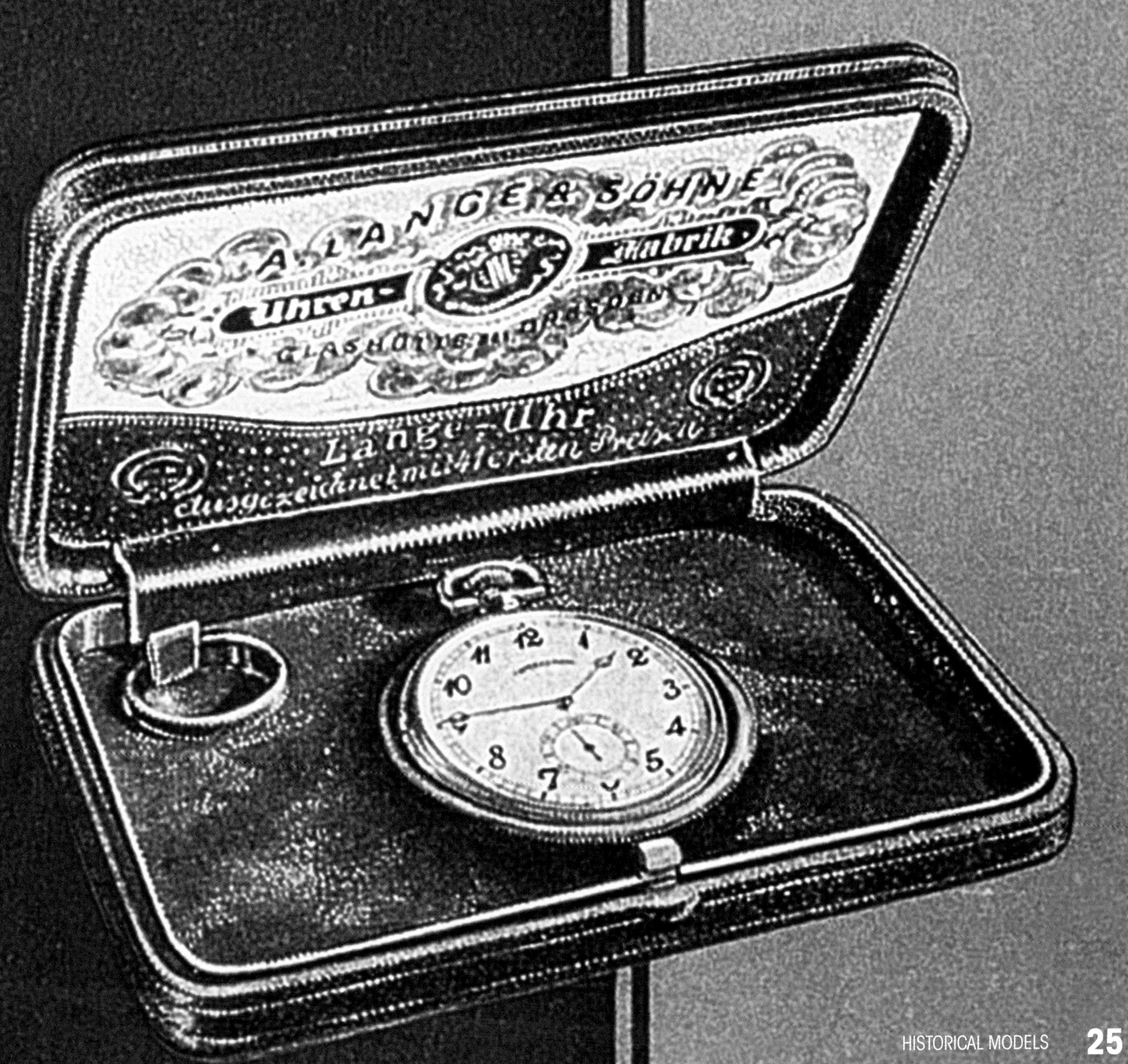
A. LANGE & SÖHNE
Uhren-
Fabrik
Lange-Uhr

A. LANGE & SÖHNE

HIGHLIGHTS

Upon its refounding in 1990, A. Lange & Söhne was very clear that the company would not just manufacture wristwatches in a reunited Germany that "simply" represent the legacy of progenitor Ferdinand Adolph Lange outfitted with typical Glashütte elements. Instead, the Tourbillon Pour le Mérite came to life in the debut collection of 1994, a timepiece representing the continuing demands of the Saxons upon themselves—as in the past the motto was clear: no more and no less than to manufacture the best mechanical watches in the world. The tourbillon, already so impressive fifteen years ago, was only the beginning of a series of exceptional timekeepers boasting the most complex technical complications, which re-established the brand in the arena of haute horlogerie (high watchmaking). To this day, the Datograph, the Lange 31, and Tourbograph Pour le Mérite, the Lange Double Split, and the Lange Zeitwerk introduced in 2009 have counted among the highlights raising the bar of traditional watchmaking.

A. LANGE & SÖHNE

LANGE ZEITWERK

"How do we plausibly combine the principles of the mechanical watch with a modern depiction of time?" This question was at the base of the development of the Lange Zeitwerk, and the answer that the Saxon watchmakers found is a clear one: a cleanly designed mechanical wristwatch with a precisely jumping digital indication. The Zeitwerk's digital display thus builds a bridge between Lange's tradition and the present without renouncing the character of the traditional Glashütte brand; it represents a true Lange timepiece for the twenty-first century.

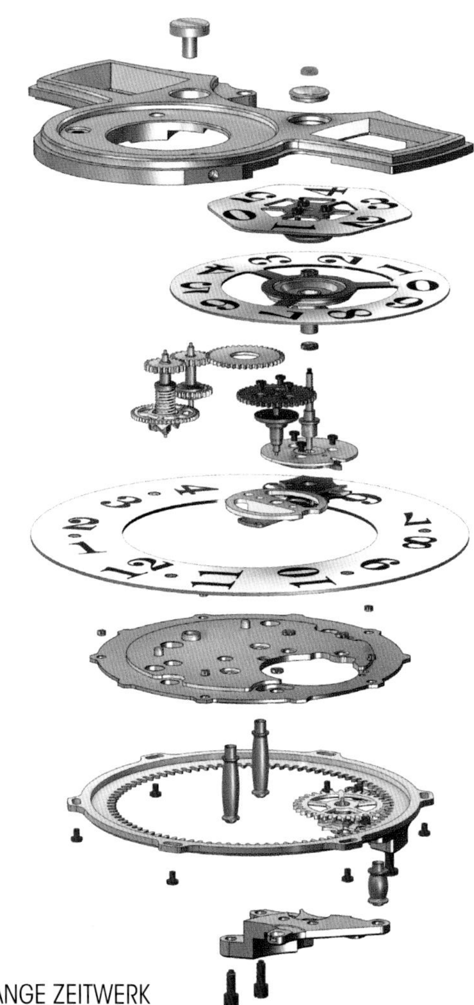

LANGE ZEITWERK

Reference number: 40.029
Movement: manual winding, Lange Caliber L043.1; hand-engraved balance cock; hand-finished components; two screw-mounted gold chatons; constant force escapement (remontoir)
Functions: hours and minutes (digital, jump), subsidiary seconds; power reserve display
Case: white gold, ø 41.9 mm, sapphire crystal; transparent case back
Price (2009): € 42.500,– [$54,600]

LANGE ZEITWERK LUMINOUS

Reference number 140.035
Movement: manual winding, Lange Caliber L043.1; hand-engraved balance cock; hand-finished components; two screw-mounted gold chatons; constant force escapement (remontoir)
Functions: hours and minutes (digital, jump), subsidiary seconds; power reserve display
Case: white gold, ø 41.9 mm, sapphire crystal; transparent case back
Price (2009): € 69.000,– [$88,600]

A.LANGE & SÖHNE TOURBILLON „POUR LE MÉRITE"

The Tourbillon Pour le Mérite was introduced in the debut collection of 1994 as its absolute highlight. The predicate "pour le mérite" refers to the Prussian order of merit for outstanding scientific achievements. In view of the fact that the world of haute horlogerie had few tourbillons to boast at the time, experts were impressed by the accomplishment of the Saxon watchmakers. Above all, it was the paper-thin chain complementing a fusee that premiered here in a wristwatch that commanded respect from connoisseurs. At the time, the watch was limited to 50 pieces in platinum and 150 pieces in gold. It has become a sought-after collector's piece since the end of the 1990s.

TECHNICAL DATA

TOURBILLON POUR LE MÉRITE

Reference number:	701.005
Movement:	manual winding, Lange Caliber L902.0; one-minute tourbillon; three-quarter plate
Functions:	hours, minutes, subsidiary seconds; power reserve display
Case:	platinum, ø 38.5 m; sapphire crystal; transparent case back
Price (1994):	€ 70.000,– [$89,900]

A.LANGE & SÖHNE DATOGRAPH

The Datograph was introduced in 1999 at the end of the previous century and set a new benchmark for the appearance and technology of a chronograph. Just like its pocket watch role models, which Ferdinand Adolph Lange was already manufacturing in 1868, this timepiece is outfitted with so-called column wheel technology, which to this day remains the luxury class of chronograph technology.

TECHNICAL DATA

DATOGRAPH

Reference number: 403.032
Movement: manual winding, Lange Caliber L951.1; hand-engraved balance cock; hand-finished components; column-wheel control of chronograph functions
Functions: hours, minutes, subsidiary seconds; chronograph with flyback function and precisely jumping minute counter, large date
Case: red gold, ø 39 mm, sapphire crystal; transparent case back
Price (2009): € 39.700,– [$51,000]

DATOGRAPH PERPETUAL

Reference number: 410.030
Movement: manual winding, Lange Caliber L952.1; hand-engraved balance cock; hand-finished components; column-wheel control of chronograph functions
Functions: hours, minutes, subsidiary seconds; chronograph with flyback function; perpetual calendar with large date, day, month, moon phase, leap year; 24-hour display with day/night indicator
Case: white gold, ø 41 mm, sapphire crystal; transparent case back
Price (2009): € 82.500,– [$106,000]

A. LANGE & SÖHNE

TOURBOGRAPH „POUR LE MÉRITE"

The Tourbograph Pour le Mérite introduced in 2005 saw A. Lange & Söhne uniting just about every outstanding horological element the brand was in a position to offer. This superlative timepiece practically constitutes a synthesis of the incredibly impressive Datograph and Tourbillon Pour le Mérite models: the limited treasure of an initial 51 pieces in platinum was the first wristwatch in the world to include a one-minute tourbillon, chain and fusee, a split-seconds chronograph, and a power reserve display.

TECHNICAL DATA

TOURBOGRAPH POUR LE MÉRITE

Reference number: 702.025

Movement: manual winding, Lange Caliber L903.0;
hand-engraved balance cock; hand-finished components;
constant force regulated by chain and fusee;
stepped planetary gear

Functions: hours, minutes, split-seconds chronograph;
power reserve display

Case: platinum, ø 41.2 mm, sapphire crystal;
transparent case back

Price (2009): upon request

A.LANGE & SÖHNE

LANGE 31

The Lange 31 was introduced in 2007 and the name suggests a special technical feature is hidden underneath its dial. With one winding this watch runs for an entire month and thus longer than any other wristwatch with one exception. Two powerful spring barrels and one winding mechanism, which—like some pocket watches—is wound using a key, allows an exceptional power reserve to be realized. Together with the constant force mechanism, which supplies the escapement with an even amount of energy regardless of the state of the mainsprings, it ensures that the same amount of energy is supplied right up to the 31st day, guaranteeing the same rate precision.

LANGE 31

Reference number: 130.032
Movement: manual winding, Lange Caliber L043.1; hand-engraved balance cock; hand-finished components; mechanically limited power reserve of 31 days; two spring barrels; constant force mechanism (remontoir)
Functions: hours, minutes, subsidiary seconds; large date; power reserve display
Case: red gold, ø 45.9 mm, sapphire crystal; transparent case back
Price (2009): € 87.000,– [$111,800]

LANGE 31

Reference number: 130.025
Movement: manual winding, Lange Caliber L043.1; hand-engraved balance cock; hand-finished components; mechanically limited power reserve of 31 days; two spring barrels; constant force mechanism (remontoir)
Functions: hours, minutes, subsidiary seconds; large date; power reserve display
Case: platinum, ø 45.9 mm, sapphire crystal; transparent case back
Price (2009): € 115.000,– [$147,700]

A. LANGE & SÖHNE
LANGE DOUBLE SPLIT

Several years of development were necessary before A. Lange & Söhne could introduce Caliber L001.1 outfitted with a double rattrapante in 2004. The Lange Double Split distinguishes itself by being the first chronograph ever to include a split function for both seconds and minutes. Thus, it is possible to measure interval times without losing sight of the entire timing process in the background. To that point, it had only been possible to measure a split-seconds interval of up to 60 seconds. Whether the precious platinum chronograph is actually used during sports events or at the racetrack remains to be seen, but the fact is that A. Lange & Söhne's Double Split has raised mechanical timekeeping to a new level.

LANGE DOUBLE SPLIT

Reference number: 404.032
Movement: manual winding, Lange Caliber L001.1; two column wheels to control chronograph's double rattrapante functions; precisely jumping minute counter; hand-engraved balance cock; hand-finished components
Functions: hours, minutes, subsidiary seconds; chronograph with flyback function; split-seconds and minute counters; power reserve display
Case: red gold, ø 43 mm, sapphire crystal; transparent case back
Price (2009): € 83.000,– [$106,600]

TECHNICAL DATA

LANGE DOUBLE SPLIT

Reference number: 404.035
Movement: manual winding, Lange Caliber L001.1; two column wheels to control chronograph's double rattrapante functions; precisely jumping minute counter; hand-engraved balance cock; hand-finished components
Functions: hours, minutes, subsidiary seconds; chronograph with flyback function; split-seconds and minute counters; power reserve display
Case: platinum, ø 43 mm, sapphire crystal; transparent case back
Price (2010): € 96.600,– [$124,100]

A.LANGE & SÖHNE
LANGE 1

Since its launch in the brand's debut collection of 1994, it would be unimaginable for the Lange 1 not to be a big part of the Saxon manufacturer's portfolio. Today, this model can justifiably be described as a classic of modern watch history. It is not only that the Lange 1, with its timeless elegance, has decisively contributed to reestablishing the modern A. Lange & Söhne brand, with its unusual dial design, but it has even ascended to become the symbol of the historically relevant brand that is today without a doubt its best-known model. These days, the Lange 1 comes outfitted with a moon phase, a second time zone, and a tourbillon among other things. It is also available in various case sizes so that watch lovers should have no trouble finding just the right personal variation.

LANGE 1

Reference number: 101.021
Movement: manual winding, Lange Caliber L901.0;
hand-engraved balance cock;
components finished and assembled by hand;
72-hour power reserve; twin spring barrels
Functions: hours, minutes, subsidiary seconds;
large date; power reserve display
Case: yellow gold, ø 38.5 mm; sapphire crystal;
transparent case back
Price (2009): € 21.500,– [$27,600]

TECHNICAL DATA

LANGE 1

Reference number: 101.039e
Movement: manual winding, Lange Caliber L901.0; hand-engraved balance cock; components finished and assembled by hand; 72-hour power reserve; twin spring barrels
Functions: hours, minutes, subsidiary seconds; large date; power reserve display
Case: white gold, ø 38.5 mm; sapphire crystal; transparent case back
Price (2009): € 21.500,– [$27,600]

LANGE 1

Reference number: 101.033
Movement: manual winding, Lange Caliber L901.0; hand-engraved balance cock; components finished and assembled by hand; 72-hour power reserve; twin spring barrels
Functions: hours, minutes, subsidiary seconds; large date; power reserve display
Case: red gold, ø 38.5 mm; sapphire crystal; transparent case back
Price (2009): € 21.500,– [$27,600]

LANGE 1

Reference number: 101.035
Movement: manual winding, Lange Caliber L901.0; hand-engraved balance cock; components finished and assembled by hand; 72-hour power reserve; twin spring barrels
Functions: hours, minutes, subsidiary seconds; large date; power reserve display
Case: platinum, ø 38.5 mm; sapphire crystal; transparent case back
Price (2009): € 31.500,– [$40,400]

LANGE 1

Reference number: 101.032
Movement: manual winding, Lange Caliber L901.0; hand-engraved balance cock; components finished and assembled by hand; 72-hour power reserve; twin spring barrels
Functions: hours, minutes, subsidiary seconds; large date; power reserve display
Case: red gold, ø 38.5 mm, height 10 mm; sapphire crystal; transparent case back
Price (2009): € 21.500,– [$27,600]

TECHNICAL DATA

GRAND LANGE 1

Reference number: 115.032
Movement: manual winding, Lange Caliber L901.2; hand-engraved balance cock; components finished and assembled by hand; 72-hour power reserve; twin spring barrels
Functions: hours, minutes, subsidiary seconds; large date; power reserve display
Case: red gold, ø 41.9 mm; sapphire crystal; transparent case back
Price (2009): € 24.000,– [$30,800]

GRAND LANGE 1 LUMINOUS

Reference number: 115.028
Movement: manual winding, Lange Caliber L901.2; hand-engraved balance cock; components finished and assembled by hand; 72-hour power reserve;twin spring barrels
Functions: hours, minutes, subsidiary seconds; large date; power reserve display
Case: white gold, ø 41.9 mm; sapphire crystal; transparent case back
Price (2010): € 24.900,– [$32,000]

GRAND LANGE 1 LUNA MUNDI/SOUTHERN CROSS

Reference number: 119.032
Movement: manual winding, Lange Caliber L901.7; moon phase display of the southern hemisphere with continuous drive via hour wheel
Functions: hours, minutes, subsidiary seconds; large date; moon phase; power reserve display
Case: red gold, ø 41.9 mm; sapphire crystal; transparent case back
Price (2009): € 49.800,– [$64,000]

GRAND LANGE 1 LUNA MUNDI/URSA MAJOR

Reference number: 119.026
Movement: manual winding, Lange Caliber L901.7; moon phase display of the northern hemisphere with continuous drive via hour wheel
Functions: hours, minutes, subsidiary seconds; large date; moon phase; power reserve display
Case: white gold, ø 41.9 mm; sapphire crystal; transparent case back
Price (2009): € 49.800,– [$64,000]

TECHNICAL DATA

LANGE 1 TIME ZONE

Reference number: 116.021

Movement: manual winding, Lange Caliber L031.1; hand-engraved balance cock; components finished by hand; home time with day/night indication, local time (hour, minute) with day/night indication and reference city ring, settable via button; 72-hour power reserve; twin spring barrels

Functions: hours, minutes, subsidiary seconds; second time zone; large date; power reserve display; day/night indication for both time zones

Case: yellow gold, ø 41.9 mm; sapphire crystal; transparent case back

Price (2009): € 30.000,– [$38,500]

LANGE 1 TIME ZONE

Reference number: 116.033

Movement: manual winding, Lange Caliber L031.1; hand-engraved balance cock; components finished by hand; home time with day/night indication, local time (hour, minute) with day/night indication and reference city ring, settable via button; 72-hour power reserve; twin spring barrels

Functions: hours, minutes, subsidiary seconds; second time zone; large date; power reserve display; day/night indication for both time zones

Case: red gold, ø 41.9 mm; sapphire crystal; transparent case back

Price (2009): € 30.000,– [$38,500]

LANGE 1 TIME ZONE

Reference number: 116.025

Movement: manual winding, Lange Caliber L031.1;
hand-engraved balance cock;
components finished by hand;
home time with day/night indication,
local time (hour, minute) with day/night
indication and reference city ring, settable via button;
72-hour power reserve; twin spring barrels

Functions: hours, minutes, subsidiary seconds; second time zone;
large date; power reserve display;
day/night indication for both time zones

Case: platinum, ø 41.9 mm; sapphire crystal;
transparent case back

Price (2009): € 41.000,– [$52,600]

TECHNICAL DATA

LANGE 1 MOONPHASE

Reference number: 109.021
Movement: manual winding, Lange Caliber L901.5; hand-engraved balance cock; components finished by hand; 72-hour power reserve; twin spring barrels; moon phase display with continuous drive via hour wheel; button for quick-set date
Functions: hours, minutes, subsidiary seconds; large date; moon phase; power reserve display
Case: yellow gold, ø 38.5 mm; sapphire crystal; transparent case back
Price (2009): € 25.600,– [$32,900]

LANGE 1 MOONPHASE

Reference number: 109.032
Movement: manual winding, Lange Caliber L901.5; hand-engraved balance cock; components finished by hand; 72-hour power reserve; twin spring barrels; moon phase display with continuous drive via hour wheel; button for quick-set date
Functions: hours, minutes, subsidiary seconds; large date; moon phase; power reserve display
Case: red gold, ø 38.5 mm; sapphire crystal; transparent case back
Price (2009): € 25.600,– [$32,900]

LANGE 1 MOONPHASE

Reference number:	109.025
Movement:	manual winding, Lange Caliber L901.5; hand-engraved balance cock; components finished by hand; 72-hour power reserve; twin spring barrels; moon phase display with continuous drive via hour wheel; button for quick-set date
Functions:	hours, minutes, subsidiary seconds; large date; moon phase; power reserve display
Case:	platinum, ø 38.5 mm; sapphire crystal; transparent case back
Price (2009):	€ 35.600,– [$45,700]

LANGE 1 TOURBILLON

Reference number: 704.025
Movement: manual winding, Lange Caliber L961.1; 72 hours power reserve; twin spring barrels; one-minute tourbillon
Functions: hours, minutes, subsidiary seconds; large date; power reserve display
Case: platinum, ø 38.5 mm; sapphire crystal; transparent case back
Price (2009): € 91.100,– [$117,000]

LANGE 1 TOURBILLON HOMAGE TO F.A. LANGE

Reference number: 722.050
Movement: manual winding, Lange Caliber L961.1; 72 hours power reserve; twin spring barrels; one-minute tourbillon
Functions: hours, minutes, subsidiary seconds; large date; power reserve display
Case: honey-colored gold, ø 38.5 mm; sapphire crystal; transparent case back
Price: € 130.000,– [$167,000]

LANGE 1 SOIRÉE

Reference number: 110.030
Movement: manual winding, Lange Caliber L901.1; 72 hours power reserve; twin spring barrels
Functions: hours, minutes, subsidiary seconds; large date; moon phase; power reserve display
Case: white gold, ø 38.5 mm; sapphire crystal; transparent case back
Price (2003): € 20.200,– [$25,900]

LITTLE LANGE 1 MOONPHASE SOIRÉE

Reference number: 819.049
Movement: manual winding, Lange Caliber L901.9; hand-engraved balance cock; components finished and assembled by hand; 72-hour power reserve; twin spring barrels
Functions: hours, minutes, subsidiary seconds; large date; moon phase; power reserve display
Case: white gold, ø 36.8 mm; bezel set with 58 brilliant-cut diamonds; transparent case back
Price (2009): € 40.400,– [$51,900]

LITTLE LANGE 1 SOIRÉE

Reference number: 813.044
Movement: manual winding, Lange Caliber L901.9; twin spring barrels
Functions: hours, minutes, subsidiary seconds; large date; moon phase; power reserve display
Case: white gold, ø 36.1 mm; bezel set with 52 brilliant-cut diamonds; sapphire crystal; transparent case back
Price (2008): € 39.500,– [$50,700]

A. LANGE & SÖHNE
RICHARD LANGE

Richard Lange, the oldest son of company founder Ferdinand Adolph Lange, understood like no other watchmaker of his era how to allow his scientific findings to flow into his movement design, guaranteeing the precision of his company's own pocket watches even under extreme conditions. The Richard Lange model reflects upon the tradition of these precise observation watches while bucking the trend of ever more numerous complications with its purist three-hand design—behind which the finest technology that Glashütte's art of horology has to offer can be found.

TECHNICAL DATA

RICHARD LANGE

Reference number: 232.025
Movement: manual winding, Lange Caliber L041.2; hand-engraved balance cock; hand-finished components
Functions: hours, minutes, sweep seconds
Case: platinum, ø 40.5 mm; sapphire crystal; transparent case back
Price (2009): € 29.800,– [$38,200]

RICHARD LANGE REFERENZUHR

Reference number: 250.032
Movement: manual winding, Lange Caliber L033.1; hand-engraved balance cock; hand-finished components
Functions: hours, minutes, subsidiary seconds, power reserve display
Case: red gold, ø 40.5 mm; sapphire crystal; transparent case back
Price (2010): € 39.500,– [$50,700]

RICHARD LANGE POUR LE MÉRITE

Reference number: 260.032
Movement: manual winding, Lange Caliber L044.1; chain and fusée; hand-engraved balance, escape wheel, and fourth wheel cocks; hand-finished components
Functions: hours, minutes, subsidiary seconds
Case: red gold, ø 40.5 mm; sapphire crystal; transparent case back
Remark: enamel dial
Price (2009): € 82.000,– [$105,300]

RICHARD LANGE POUR LE MÉRITE

Reference number: 206.025
Movement: manual winding, Lange Caliber L044.1; chain and fusée; hand-engraved balance, escape wheel, and fourth wheel cocks; hand-finished components
Functions: hours, minutes, subsidiary seconds
Case: red gold, ø 40.5 mm; sapphire crystal; transparent case back
Price (2009): € 98.000,– [$125,900]

A. LANGE & SÖHNE 1815

The year 1815 remains a very special one for A. Lange & Söhne: while the reorganization of Europe was being disputed at the Congress of Vienna, eventual company founder Ferdinand Adolph Lange was born in Dresden. In 1995, precisely 180 years later, the reborn brand A. Lange & Söhne reminisced on the event with a new model called 1815. This fifth new development of the new-old brand was presented as a classic men's piece featuring a design reduced to the purist's minimum. Its simple beauty has been a collection staple of the masterful Glashütte watchmakers for the last fifteen years.

1815

Reference number: 206.025
Movement: manual winding, Lange Caliber L941.1
Functions: hours, minutes, subsidiary seconds
Case: platinum, ø 35.9 mm; sapphire crystal; transparent case back
Price (2006): € 13.600,–

1815

Reference number: 206.032
Movement: manual winding, Lange Caliber L941.1
Functions: hours, minutes, subsidiary seconds
Case: red gold, ø 35.9 mm; sapphire crystal; transparent case back
Price (2006): € 8300,– [$10,600]

1815

Reference number:	233.026
Movement:	manual winding, Lange Caliber L051.1; hand-engraved balance cock; components finished and assembled by hand; 55 hours power reserve
Functions:	hours, minutes, subsidiary seconds
Case:	white gold, ø 40 mm; sapphire crystal; transparent case back
Price (2009):	€ 14.300,– [$18,300]

1815

Reference number:	233.021
Movement:	manual winding, Lange Caliber L051.1; hand-engraved balance cock; components finished and assembled by hand; 55 hours power reserve
Functions:	hours, minutes, subsidiary seconds
Case:	yellow gold, ø 40 mm; sapphire crystal; transparent case back
Band:	reptile skin, buckle
Price (2009):	€ 14.300,– [$18,300]

TECHNICAL DATA

1815 UP AND DOWN

Reference number: 221.032
Movement: manual winding, Lange Caliber L942.1
Functions: hours, minutes, subsidiary seconds, power reserve display
Case: red gold, ø 35.9 mm; sapphire crystal; transparent case back
Price (2006): € 12.000,– [$15,400]

1815 UP AND DOWN

Reference number: 21.025
Movement: manual winding, Lange Caliber L942.1
Functions: hours, minutes, subsidiary seconds; power reserve display
Case: platinum, ø 35.9 mm; sapphire crystal; transparent case back
Price (2006): € 17.200,– [$22,100]

1815 MOONPHASE

Reference number: 231.035
Movement: manual winding, Lange Caliber L943.1; swan-neck fine adjustment; three-quarter plate
Functions: hours, minutes, subsidiary seconds; moon phase
Case: platinum, ø 35.9 mm; sapphire crystal; transparent case back
Price (2000): € 13.900,– [$17,800]

1815 MOONPHASE HOMAGE TO F.A. LANGE

Reference number: 212.050
Movement: manual winding, Lange Caliber L943.1; swan-neck fine adjustment; three-quarter plate
Functions: hours, minutes, subsidiary seconds; moon phase
Case: honey-colored gold, ø 35.9 mm; sapphire crystal; transparent case back
Price (2010): € 18.500,– [$23,700]

1815 AUTOMATIC

Reference number: 303.021
Movement: automatic, Lange Caliber L921.2
Functions: hours, minutes, subsidiary seconds
Case: yellow gold, ø 37 mm; sapphire crystal; transparent case back
Price (2006): € 12.900,– [$16,500]

1815 CHRONOGRAPH

Reference number: 402.026
Movement: manual winding, Lange Caliber L951.5; column-wheel control of chronograph functions
Functions: hours, minutes, subsidiary seconds; chronograph with flyback function
Case: white gold, ø 39.5 mm; sapphire crystal; transparent case back
Price (2010): € 32.500,– [$41,700]

1815 CHRONOGRAPH

Reference number: 401.026
Movement: manual winding, Lange Caliber L951.0; column-wheel control of chronograph functions
Functions: hours, minutes, subsidiary seconds; chronograph with flyback function
Case: red gold, ø 39.5 mm; sapphire crystal; transparent case back
Price (2004): € 29.600,– [$38,000]

A. LANGE & SÖHNE
CABARET

It is a clean, round shape that characterizes almost all of A. Lange & Söhne's timepieces—but only almost. In contrast to these, the Cabaret, with its rectangular shape, attracts a spotlight all its own and offers those who love this classic case shape a true Lange timepiece. The striking large date, the dial design reduced to the minimum with its reserved markers, and the easily recognizable subsidiary seconds offer everything that the brand's round watches do. The Cabaret in all of its variations including moon phase and tourbillon is no less a Lange than its round siblings.

TECHNICAL DATA

CABARET

Reference number: 107.031
Movement: manual winding, Lange Caliber L931.3; hand-engraved balance cock
Functions: hours, minutes, subsidiary seconds; large date
Case: red gold, 36.3 x 25.5 mm; sapphire crystal; transparent case back
Price (2009): € 17.000,– [$21,800]

TECHNICAL DATA

CABARET

Reference number: 107.021
Movement: manual winding, Lange Caliber L931.3; hand-engraved balance cock
Functions: hours, minutes, subsidiary seconds; large date
Case: yellow gold, 36.3 x 25.5 mm; sapphire crystal; transparent case back
Price (2009): € 17.000,– [$21,800]

CABARET

Reference number: 107.031
Movement: manual winding, Lange Caliber L931.3; hand-engraved balance cock
Functions: hours, minutes, subsidiary seconds; large date
Case: platinum, 36.3 x 25.5 mm; sapphire crystal; transparent case back
Price (2000): € 18.000,– [$23,100]

CABARET

Reference number: 157.127
Movement: manual winding, Lange Caliber L931.3; hand-engraved balance cock
Functions: hours, minutes, subsidiary seconds; large date
Case: white gold, 36.3 x 25.5 mm; sapphire crystal; transparent case back
Price (2006): € 25.100,– [$32,200]

CABARET

Reference number: 157.132
Movement: manual winding, Lange Caliber L931.3; hand-engraved balance cock
Functions: hours, minutes, subsidiary seconds; large date
Case: red gold, 36.3 x 25.5 mm; sapphire crystal; transparent case back
Price (2006): € 25.100,– [$32,200]

TECHNICAL DATA

CABARET MOONPHASE

Reference number: 118.021
Movement: manual winding, Lange Caliber L931.5; hand-engraved balance cock
Functions: hours, minutes, subsidiary seconds; large date; moon phase
Case: yellow gold, 36.3 x 25.5 mm; sapphire crystal; transparent case back
Price (2009): € 20.100,– [$25,800]

CABARET TOURBILLON

Reference number: 703.025
Movement: manual winding, Lange Caliber L931.4; one-minute tourbillon with stop-seconds, hand-engraved intermediate wheel and tourbillon bridge; 5 days power reserve
Functions: hours, minutes, subsidiary seconds; large date; power reserve display
Case: platinum, 39.2 x 29.5 mm; sapphire crystal; transparent case back
Price (2009): € 205.000,– [$263,400]

CABARET TOURBILLON

Reference number: 703.032
Movement: manual winding, Lange Caliber L931.4; one-minute tourbillon with stop-seconds, hand-engraved intermediate wheel and tourbillon bridge; 5 days power reserve
Functions: hours, minutes, subsidiary seconds; large date; power reserve display
Case: red gold, 39.2 x 29.5 mm; sapphire crystal; transparent case back
Price (2009): € 182.000,–[$233,900]

TECHNICAL DATA

CABARET JEWELRY

Reference number: 808.040
Movement: manual winding, Lange Caliber L931.3
Functions: hours, minutes, subsidiary seconds; large date
Case: red gold, 36.3 x 26.5 mm;
bezel and lugs set with brilliant-cut diamonds;
sapphire crystal; transparent case back
Price (2006): € 34.200,– [$43,900]

CABARET JEWELRY

Reference number: 808.033
Movement: manual winding, Lange Caliber L931.3
Functions: hours, minutes, subsidiary seconds; large date
Case: white gold, 36.3 x 25.5 mm;
case set with 244 brilliant-cut diamonds; sapphire crystal;
transparent case back; dial set with 146 diamonds
Price (2004): € 52.800,– [$67,800]

CABARET SOIRÉE

Reference number: 827.043
Movement: manual winding, Lange Caliber L931.3
Functions: hours, minutes, subsidiary seconds; large date
Case: red gold, 36.3 x 25.5 mm;
bezel set with 70 brilliant-cut diamonds;
sapphire crystal; transparent case back
Price (2007): € 29.500,– [$37,900]

A.LANGE & SÖHNE
SAXONIA

The year 1815 remains a very special one for A. Lange & Söhne: while the reorganization of Europe was being disputed at the Congress of Vienna, eventual company founder Ferdinand Adolph Lange was born in Dresden. In 1995, precisely 180 years later, the reborn brand A. Lange & Söhne reminisced on the event with a new model called 1815. This fifth new development of the new-old brand was presented as a classic men's piece featuring a design reduced to the purist's minimum. Its simple beauty has been a collection staple of the masterful Glashütte watchmakers for the last fifteen years.

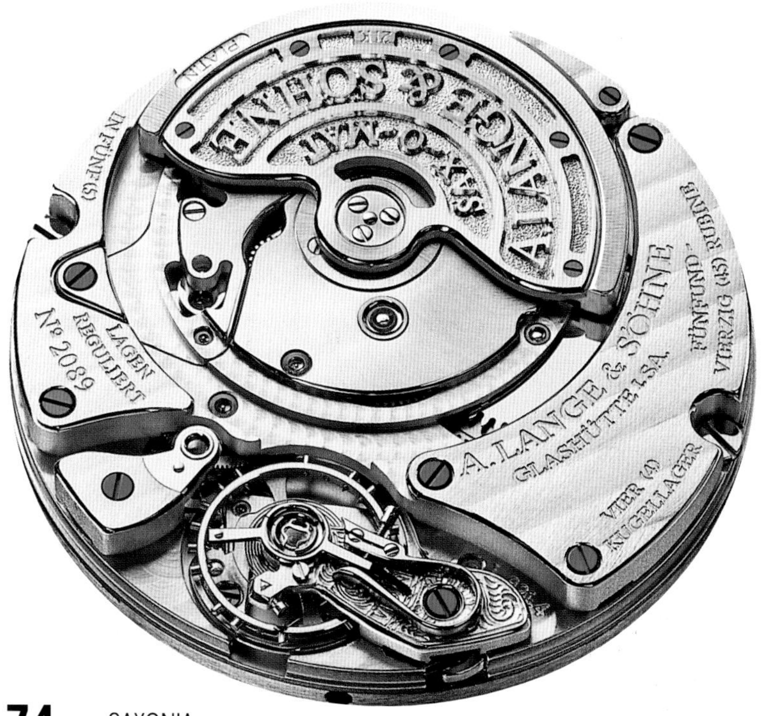

TECHNICAL DATA

SAXONIA

Reference number:	215.021
Movement:	manual winding; Lange Caliber L941.1; hand-engraved balance cock; hand-finished components
Functions:	hours, minutes, subsidiary seconds
Case:	yellow gold, ø 37 mm; sapphire crystal; transparent case back
Price (2009):	€ 12.900,– [$16,500]

TECHNICAL DATA

SAXONIA

Reference number: 105.025
Movement: manual winding; Lange Caliber L941.3; hand-engraved balance cock; hand-finished components
Functions: hours, minutes, subsidiary seconds; large date
Case: platinum, ø 33.9 mm; sapphire crystal; transparent case back
Price (2006): € 18.500,– [$23,700]

SAXONIA ANNUAL CALENDAR

Reference number: 330.026
Movement: automatic; Lange Caliber L085.1 SAX-O-MAT; three-quarter plate; hand-engraved balance cock; hand-finished components
Functions: hours, minutes, subsidiary seconds; annual calendar with large date, day of the week, month, and moon phase
Case: white gold, ø 38.5 mm; sapphire crystal; transparent case back
Price (2010): € 27.500,– [$35,300]

SAXONIA AUTOMATIC

Reference number: 315.032
Movement: automatic; Lange Caliber L921.4 SAX-O-MAT; hand-engraved balance cock; hand-finished components; Zero Reset hand-setting mechanism
Functions: hours, minutes, subsidiary seconds; large date
Case: red gold, ø 37 mm; sapphire crystal; transparent case back
Price (2009): € 20.600,– [$26,400]

SAXONIA AUTOMATIC

Reference number: 315.033
Movement: automatic; Lange Caliber L921.4 SAX-O-MAT; hand-engraved balance cock; hand-finished components; Zero Reset hand-setting mechanism
Functions: hours, minutes, subsidiary seconds; large date
Case: red gold, ø 37 mm; sapphire crystal; transparent case back
Price (2009): € 20.600,– [$26,400]

TECHNICAL DATA

GRAND SAXONIA AUTOMATIC

Reference number: 307.026
Movement: automatic; Lange Caliber L921.2 SAX-O-MAT; hand-engraved balance cock; hand-finished components; Zero Reset hand-setting mechanism
Functions: hours, minutes, subsidiary seconds
Case: white gold, ø 40 mm; sapphire crystal; transparent case back
Price (2009): € 18.300,– [$23,500]

GRAND SAXONIA AUTOMATIC

Reference number: 307.029
Movement: automatic; Lange Caliber L921.2 SAX-O-MAT; hand-engraved balance cock; hand-finished components; Zero Reset hand-setting mechanism
Functions: hours, minutes, subsidiary seconds
Case: white gold, ø 40 mm; sapphire crystal; transparent case back
Price (2009): € 18.300,– [$23,500]

SAXONIA SOIRÉE

Reference number: 835.021
Movement: manual winding; Lange Caliber L941.2; three-quarter plate; hand-engraved balance cock; hand-finished components
Functions: hours, minutes, subsidiary seconds
Case: yellow gold, ø 34 mm; bezel set with 52 brilliant-cut diamonds; sapphire crystal; transparent case back
Price (2010): € 24.500,– [$31,400]

A LANGE & SÖHNE
LANGEMATIK

Classically designed wristwatches generally dispose of manual winding to tension the mainspring, even if not every person loves this forced relationship to one's watch. Thus, A. Lange & Söhne presented the SAX-O-MAT automatic movement in 1997, three years after the brand's return to the world of fine mechanics. Its name alone displays the company's pride in its history and geographical origins. Emil and Richard Lange, the sons of the company founder, had already developed a pocket watch with automatic movement in 1891, and thus the Langematik can well be considered a belated progression of this principle that satisfies those who do not want their Lange timepieces to ever stop running.

LANGEMATIK

Reference number: 308.021
Movement: automatic, Lange Caliber L921.4 SAX-O-MAT; Zero Reset hand-setting mechanism
Functions: hours, minutes, subsidiary seconds; large date
Case: yellow gold, ø 37 mm; sapphire crystal; transparent case back
Price (2002): € 16.800,– [$21,500]

LANGEMATIK

Reference number: 308.027
Movement: automatic, Lange Caliber L921.4 SAX-O-MAT; Zero Reset hand-setting mechanism
Functions: hours, minutes, subsidiary seconds; large date
Case: white gold, ø 37 mm; sapphire crystal; transparent case back
Price (2006): € 18.400,– [$23,600]

LANGEMATIK

Reference number: 308.025
Movement: automatic, Lange Caliber L921.4 SAX-O-MAT; Zero Reset hand-setting mechanism
Functions: hours, minutes, subsidiary seconds; large date
Case: platinum, ø 37 mm; sapphire crystal; transparent case back
Price (2002): € 24.200,– [$31,100]

ANNIVERSARY LANGEMATIK

Reference number:	302.025
Movement:	automatic, Lange Caliber L921.7 SAX-O-MAT; Zero Reset hand-setting mechanism
Functions:	hours, minutes, subsidiary seconds
Case:	platinum, ø 37 mm; sapphire crystal; transparent case back
Price (2002):	€ 20.100,– [$25,800]

TECHNICAL DATA

GRAND LANGEMATIK

Reference number: 309.031
Movement: automatic, Lange Caliber L921.4 SAX-O-MAT; Zero Reset hand-setting mechanism
Functions: hours, minutes, subsidiary seconds; large date
Case: red gold, ø 40 mm; sapphire crystal; transparent case back
Price (2005): € 18.600,– [$23,900]

LANGEMATIK PERPETUAL

Reference number: 310.232

Movement: automatic, Lange Caliber L922.1 SAX-O-MAT;
hand-engraved balance cock; hand-finished components;
Zero Reset hand-setting mechanism;
main corrector for synchronizing all calendar functions,
additionally threeindividual ones

Functions: hours, minutes, subsidiary seconds;
perpetual calendar with large date, day, month,
moon phase, leap year; 24-hour display

Case: red gold, ø 38.5 mm; sapphire crystal;
transparent case back

Price (2009): € 50.700,– [$65,100]

LANGEMATIK PERPETUAL

Reference number: 310.225
Movement: automatic, Lange Caliber L922.1 SAX-O-MAT; hand-engraved balance cock; hand-finished components; Zero Reset hand-setting mechanism; main corrector for synchronizing all calendar functions, additionally three individual ones
Functions: hours, minutes, subsidiary seconds; perpetual calendar with large date, day, month, moon phase, leap year; 24-hour display
Case: platinum, ø 38.5 mm; sapphire crystal; transparent case back
Price (2009): € 93.800,– [$120,500]

LANGEMATIK PERPETUAL

Reference number:	310.221
Movement:	automatic, Lange Caliber L922.1 SAX-O-MAT; hand-engraved balance cock; hand-finished components; Zero Reset hand-setting mechanism; main corrector for synchronizing all calendar functions, additionally three individual ones
Functions:	hours, minutes, subsidiary seconds; perpetual calendar with large date, day, month, moon phase, leap year; 24-hour display
Case:	yellow gold, ø 38.5 mm; sapphire crystal; transparent case back
Price (2006):	€ 52.900,– [$67,900]

A. LANGE & SÖHNE
ARKADE

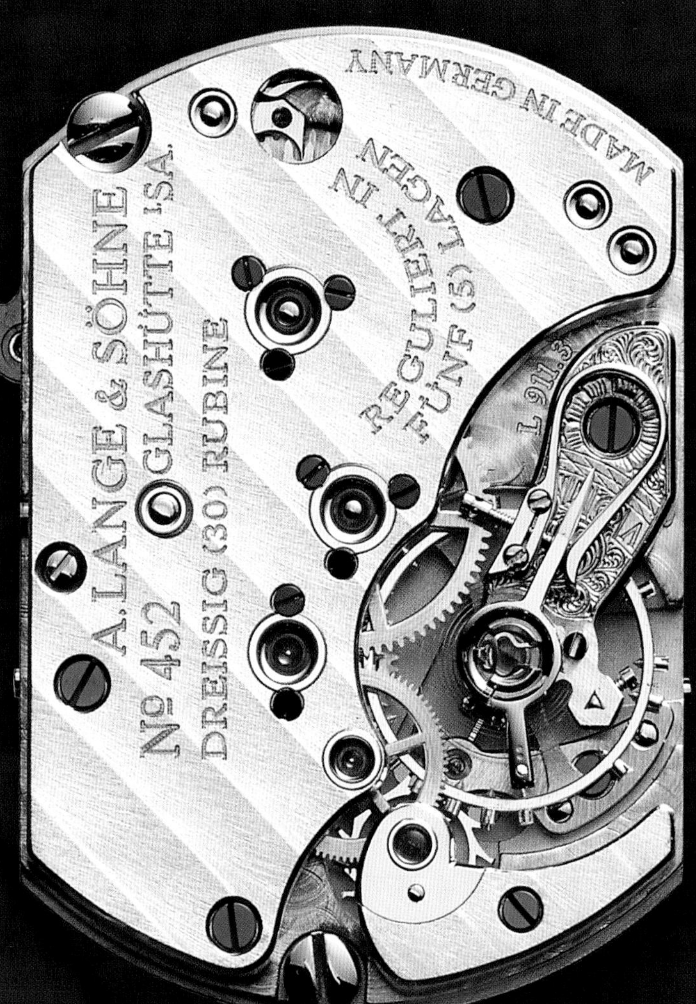

In the debut collection of 1994, A. Lange & Söhne also presented a watch for ladies: the exceptional case shape of the Arkade model was inspired by the colonnade of the Dresden Castle. Combined with its typically A. Lange & Söhne clean shape and striking large date, it unites references to the brand's Saxon home and its more than 150-year-old company history. The Arkade model is today available in a myriad of variations. Even as a simple and graceful piece of jewelry in a precious platinum case or in obvious yellow gold set with gems, they all have one thing in common: a portrayal of Saxony's legendary luster.

ARKADE

Reference number:	103.021
Movement:	manual winding, Lange Caliber L911.4; swan-neck fine adjustment; three-quarter plate
Functions:	hours, minutes, subsidiary seconds; large date
Case:	yellow gold, 29 x 22.2 mm; sapphire crystal; transparent cast back; corrector for date display
Price (2008):	€ 12.000,– [$15,400]

ARKADE

Reference number: 103.035
Movement: manual winding, Lange Caliber L911.4; swan-neck fine adjustment; three-quarter plate
Functions: hours, minutes, subsidiary seconds; large date
Case: platinum, 29 x 22.2 mm; sapphire crystal; transparent cast back; corrector for date display
Price (2006): € 20.400,– [$26,200]

ARKADE

Reference number: 153.022
Movement: manual winding, Lange Caliber L911.4; swan-neck fine adjustment; three-quarter plate
Functions: hours, minutes, subsidiary seconds; large date
Case: yellow gold, 29 x 22.2 mm; sapphire crystal; transparent cast back; corrector for date display
Price (1997): € 14.200,– [$18,200]

ARKADE

Reference number: 801.021
Movement: manual winding, Lange Caliber L911.4; swan-neck fine adjustment; three-quarter plate
Functions: hours, minutes, subsidiary seconds; large date
Case: yellow gold, 29 x 22.2 mm; bezel set with 40 brilliant-cut diamonds; sapphire crystal; transparent cast back
Price (2006): € 19.100,– [$24,500]

GRAND ARKADE

Reference number: 106.025
Movement: manual winding, Lange Caliber L911.4; swan-neck fine adjustment; three-quarter plate
Functions: hours, minutes, subsidiary seconds; large date
Case: platinum, 38 x 29.5 mm; sapphire crystal; transparent cast back
Price (2003): € 20.600,– [$26,400]

GRAND ARKADE

Reference number: 106.032
Movement: manual winding, Lange Caliber L911.4; swan-neck fine adjustment; three-quarter plate
Functions: hours, minutes, subsidiary seconds; large date
Case: red gold, 38 x 29.5 mm; sapphire crystal; transparent cast back
Price (2002): € 12.700,– [$16,300]

GRAND ARKADE

Reference number:	812.021
Movement:	manual winding, Lange Caliber L911.4; swan-neck fine adjustment; three-quarter plate
Functions:	hours, minutes, subsidiary seconds; large date
Case:	yellow gold, 38 x 29.5 mm; case set with 62 brilliant-cut diamonds; sapphire crystal; transparent cast back
Price (2005):	€ 24.000,– [$30,800]

ARKADE

Reference number: 861.031
Movement: manual winding, Lange Caliber L911.4; swan-neck fine adjustment; three-quarter plate
Functions: hours, minutes, subsidiary seconds; large date
Case: white gold, 31.5 x 24.9 mm; case set with 219 brilliant-cut diamonds; sapphire crystal; transparent cast back
Price (2005): € 128.600,– [$165,200]

TECHNICAL DATA

ARKADE

Reference number:	801.030
Movement:	manual winding, Lange Caliber L911.4; swan-neck fine adjustment; three-quarter plate
Functions:	hours, minutes, subsidiary seconds; large date
Case:	yellow gold, 29 x 22.2 mm; bezel set with 38 baguette-cut diamonds; sapphire crystal; transparent cast back
Price (2002):	€ 37.400,– [$48,000]

A. LANGE & SÖHNE®

GE & SÖHNE
ASHÜTTE I/SA
MADE IN GERMANY
A. LA